D1824214

A BOOK OF BIRDS
BY CARTON MOORE PARK

BLACKIE AND SON LIMITED
50 OLD BAILEY, LONDON, E.C.
AND GLASGOW AND DUBLIN
1900

The Flamingo

The Flamingo is most happy standing on one leg in a foot or two of water. There he waits patiently for any fish that may come his way. His colour may be pink or scarlet according to the part of the world in which he is born; and when he is standing motionless on the look-out for fish, his red body and long legs give him quite a military appearance—like a soldier at attention. The Flamingo prefers a warm climate, and by moving from place to place he manages to enjoy a continual summer-time.

The Vulture

The Vulture is a very useful bird, but he would not make a pleasant pet. His home is in the sunny lands of the south, where he is always very busy in tidying up for Dame Nature. When any poor animal is killed or dies of old age the news seems to spread like magic, for although not a single bird may be in sight, in the course of a few minutes Vultures come flocking up from all sides. And they never leave their banquet until they have eaten up everything but the bones.

The Magpie

The Magpie is not, as a rule, on good terms with his neighbours. He is very noisy, very mischievous, and very quarrelsome, and is not above stealing eggs from the nests of other birds. No doubt he clears the fields of a great number of grubs and slugs, but he does so much damage in the poultry-yard that he always goes the other way when he sees the farmer coming. His nest is built very cunningly of sticks and clay, and he surrounds it with sharp thorny twigs to keep out robbers like himself.

The Penguin

The Penguin is a kind of humpty-dumpty bird. He is far too fat to fly—the best he can do is to waddle. But his fat is very useful to him, for it is a kind of greatcoat, and helps to keep him warm while fishing in the bitterly cold waters around the South Pole. There is only one time, in fact, when the Penguin becomes lean, and that is when Mother and Father are bringing up their family. Then they seem to forget all about themselves, and while the little Penguins grow sleek and podgy, the poor old birds become so scraggy that their best friends scarcely know them.

The Wren

When the nightingale, the swallow, and many other of our summer visitors desert us, the little Wren remains through the long winter to cheer us with her song. She is a tiny bird, but her song is very loud, and sweet, and clear, and she may often be heard singing gaily even while the snow is falling. In the springtime the Wren builds her nest in the woods, choosing a place as near to the ground as possible, but later in the year she leaves her snug home and takes up her quarters under the eaves of houses, or as near as she can get to the dwellings of man.

The Eagle

The Eagle has long been regarded as the King of Birds, just as the lion is spoken of as the King of Beasts. There are some who say that he is not worthy of this honour, but certainly few of his subjects would care to fight him for the crown, for he has a remarkably strong beak, and his feet are armed with the sharpest of claws. Soaring high up in the air, he swoops down on his prey like a thunderbolt from the sky, and carries it off to his nest on some rocky cliff or steep mountain side. Small birds, rabbits, lambs, and fawns all help to fill the Eagle's larder.

The White Heron

The White Heron lives upon what he can pick up on the borders of marshes or the banks of rivers. He has a pretty large bill of fare, including fish, frogs, rats, mice, and various kinds of snails, worms, and insects. With his long legs he wades for some distance into the water, and there he stands without a movement, waiting for the fish to come within reach of his terrible beak. But at the same time he keeps a watchful eye upon what is going on around him; for the Heron is very shy, and if you go to call upon him, you will probably find that he is not at home.

The Turkey

Although his gobble-gobble is now to be heard in every farmyard, it is only a few hundred years since the Turkey first came to this country. He did not come from Turkey, as people at one time thought, but from the Continent of North America, where he is still to be found in his wild state. Mr. and Mrs. Turkey are not always on good terms, for when Mrs. Turkey lays an egg she has to hide it away to prevent her husband from destroying it. In fact, he is a tyrant, and is not afraid of anybody, except Father Christmas.

The Bird of Paradise

The Bird of Paradise is the most gorgeous of living creatures. From the crest of his head to the last feather in his magnificent tail he is a blaze of brilliant colour. And very proud he is of his good looks—so much so that a number of these beautiful birds will assemble together merely in order to show themselves off to one another. In New Guinea, where the Bird of Paradise lives, the natives call these gatherings dancing parties; but the poor hen bird, who is not at all good-looking, is not invited.

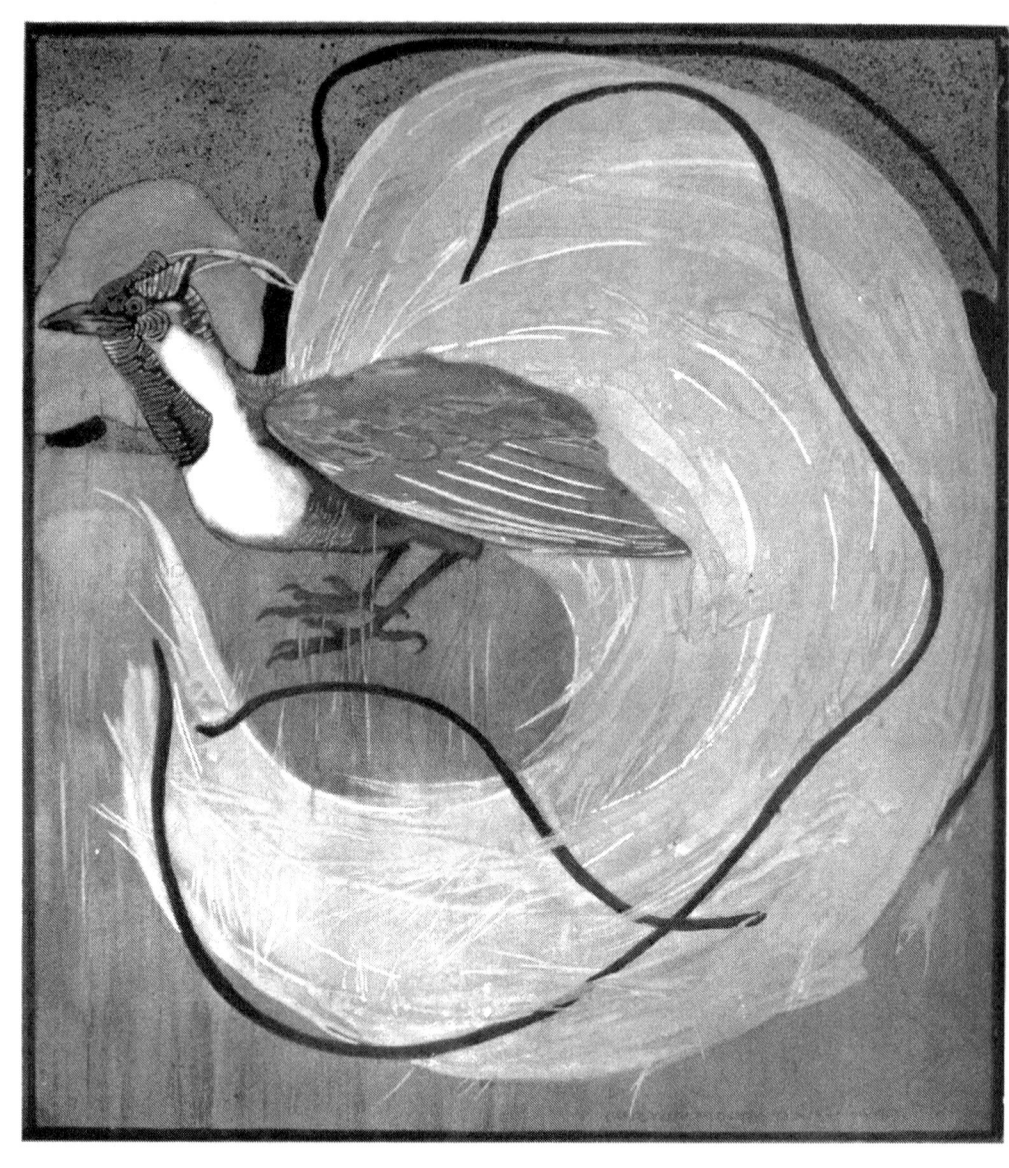

The Barn-door Fowl

The Barn-door Fowl does not often live to a great age, but her life, although short, is a very merry one. She is provided with a comfortable home, in which she can cackle to her heart's content, without fear of being snapped up by her enemy the fox; and every day she receives an ample supply of corn, to say nothing of worms, cabbage-stumps, cold potatoes, and other luxuries. In return for all this she is only asked to lay a fair number of eggs for our breakfast. If she will not do this, of course the consequences are serious.

The Adjutant

The Adjutant bird owes his name to his very dignified walk, which is believed to be almost as important as that of a real adjutant on parade. He is nearly as tall, too, as a British soldier, helmet and all, and has an even better appetite, for he can swallow a fowl or rabbit, or even a small leg of mutton, at a single mouthful. The Adjutant lives in India, where he helps the jackal and the crow to eat up what no other bird or beast will touch. When he is about, it is best to keep the larder door locked.

The Raven

The Raven is as black as a chimney-sweep, and very wicked besides. There is nothing small or weak that he will not attack; but he is particularly fond of ducklings, chickens, and young lambs. He is so knowing, too, that in olden times he was supposed to be able to foretell the future. In those days priests were specially appointed to study his croakings and tell the people what was going to happen; but this was before there were any newspapers.

The Robin Redbreast

When snow is on the ground, and King Frost holds the woods and fields in his icy grip, the little Robin Redbreast taps at our window for his breakfast of crumbs. If we are very quiet and gentle, and are careful to shut up the cat, he will even hop into the room and help himself to the good things on the table. For of all the little birds that make the woods glad with their song, there is none so fearless as this tiny warbler with the red breast, the bold black eye, and pretty winning ways.

The Condor

The Condor is seen at his best far up in the mountains of South America, where he builds his nest on a crag a hundred times as high as a church steeple. Although he is one of the largest birds of prey, he is usually content to wait until the puma has finished a meal, when he sails down and eats up what remains. But sometimes he kills a lamb or a goat on his own account, and when he finds a meal to his liking he will gorge himself until he is unable to fly. Then he is often caught and pays the penalty of his greediness.

The Goose

The ordinary farmyard Goose has only one business in life, and that is to grow fat. The farmer does not teach her any other accomplishment, so it is scarcely surprising that she is rather stupid. That the Goose can be clever if she has the chance we see from the behaviour of her untamed sisters, who are among the wariest and most intelligent of birds. They live in huge flocks, and when they settle down to feed there is always a sentry Goose on guard to warn them of the approach of danger.

The Cassowary

The Cassowary may be said to be only half a bird, for though he has wings he is quite unable to fly. But with his long legs he can run along the ground at an amazing speed, and it must be a swift horse indeed that can overtake him. His home is in the beautiful islands of the South Seas, where he finds a plentiful supply of fruit. But he is also fond of the eggs of other birds, and to assist his digestion he will swallow any scraps of old iron or broken bottles that may be lying about.

The Pelican

The Pelican is a fisherman by trade, and his fishing basket is a part of himself. Just under his bill is a large pouch in which he stores up all the fish he catches until it is time for dinner. When mamma goes home to the little Pelicans, she opens her beak and allows them to help themselves out of this basket. Let us hope that she has had her own dinner first, for they are greedy little rascals.

The Pigeon

The Pigeon is one of the gentlest and most trustful of pets. Few birds can compare with him in swiftness of flight, but although he may sometimes lose himself, it never occurs to the tame Pigeon to fly away from his dove-cote, and seek his fortune in the wide world. Like the house cat, he thinks there is no place like home. The Carrier, one of the swiftest of the Pigeon family, will find his way home over hundreds of miles, travelling faster than many an express train. In fact, the Carrier Pigeon is often used to carry messages from one distant place to another.

The Guinea Fowl

The Guinea Fowl has been so long among us that we no longer look upon him as a stranger. He was brought a long time ago from Africa, where his relations may still be found, assembling in huge flocks in damp, marshy places that furnish a good supply of worms and insects. Whether he be tame or wild, the most remarkable thing about the Guinea Fowl is his voice. When he is frightened or angry he gives forth a screech like an old barn-door creaking on rusty hinges, and he does not leave off until he and everybody else are quite tired.

The Jackdaw

The Jackdaw is a cousin of the Rook, and, like him, lives in flocks. He makes his home high up in church steeples and old, ruined towers, where he spends a great part of the day chattering and quarrelling. He quickly makes friends with sheep, and may often be seen in the fields plucking wool from their backs to line his nest with. He is easily tamed if he is caught when just learning to fly, and, as he may be taught to speak, he makes a very interesting pet.

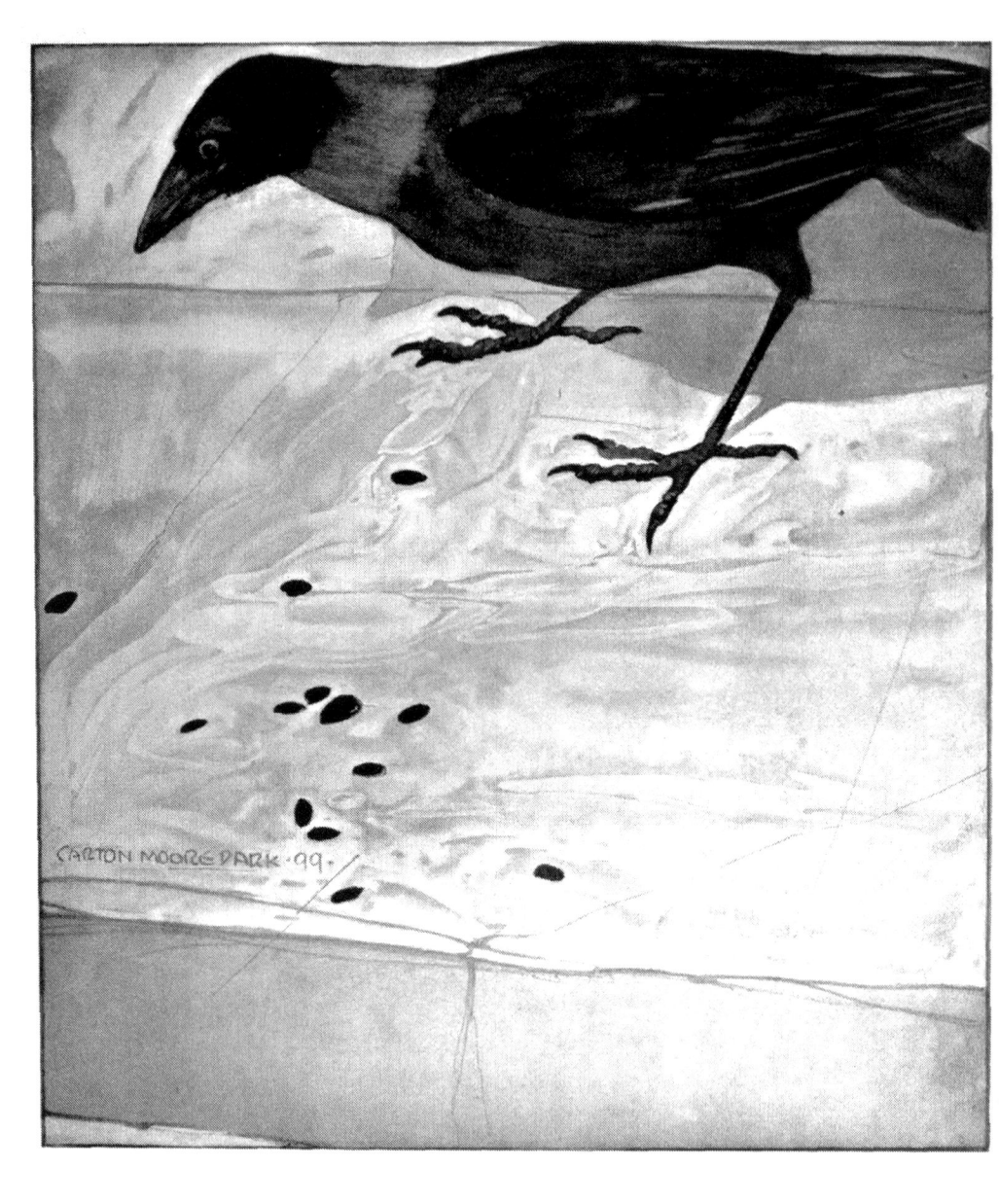

The Duck

The tame Duck, although now in humble circum-
stances, comes of a most respectable family. She is
first cousin to the wild duck, who is much sought
after at certain seasons, and among her more distant
relations are the lordly swan and the graceful flamingo.
As a swimmer and diver the tame Duck has very few
equals among feathered folk. Even as a duckling
she does not require a single swimming-lesson, but at
the first sight of water plunges boldly in and begins
hunting for worms and other delicacies.

The Peacock

On his head the Peacock carries a crest of twenty-four beautiful feathers, and behind him a train more gorgeous than that of any princess. When he is pleased he lifts up his train and spreads it out like a fan—a fan of such beautiful colours and so delightful a pattern that it could not be made for a king's ransom. In the moulting season these feathers drop off, and then the Peacock is so much ashamed of himself that he hides away until they grow again. His wife is not so richly dressed; indeed, the poor thing is quite a dowdy person.

The Seagull

When circling between sea and sky, or skimming lightly over the crests of the waves, the Seagull is a picture of beauty and grace. But all the while he has a keen eye to business; and a sudden dive, a splash, and the gleam of silvery scales tell us that another little fish will swim no more in the deep blue sea. Like Jack Tar, the Seagull gets his living on the ocean; but when fish are scarce, or the weather at sea is more than usually cold, he makes his way inland, and is content with worms and slugs and almost anything else that is eatable.

The Parrot

Until he is caught, put in a cage, and taught to say "Pretty Polly", the Parrot leads a very pleasant life. His home is usually in the very hot regions of the earth, where he makes a pretty picture with his bright plumage, flitting about in the dense forests with scores and hundreds of his friends. He lives upon fruit and honey, and when he is not feeding he is chattering and screeching. Even if his neighbour is pounced upon by a tree-snake or a four-footed enemy his grief and alarm only last for a few minutes. One parrot is never missed among so many.

The Rook

The Rook is a busy, chattering, cheerful soul, who loves plenty of noise and bustle, and is never content with his own company. In order to have his friends and relations around him, he builds his nest in a kind of bird-village, or rookery as it is called, high up in a clump of tall trees. The rookery is governed by strict laws, and one of the strictest is that strangers are not admitted on any account. If any rash new-comer ventures to begin nest-building, the old inhabitants set upon him with beak and claw, drive him out of the rookery, and tear his house to pieces.

The Owl

Nobody could be half so wise as the Owl looks; but there is no reason to suppose that he has more brains than the rest of us. By day he keeps himself to himself, for the sun is bad for his eyes; but at dusk he comes out from his hole in the belfry tower or ivy-covered wall, and flits about the fields on the look-out for his supper. When they hear his grim "Hoot-toot!" the rat, and the mole, and the little field-mouse had better hurry home to their nests.

TWENTY-SIX BIRDS

Ingram Content Group UK Ltd.
Milton Keynes UK
UKHW051824190623
423468UK00023B/35